Francileide Martins P. de Sá Leitão

Statistics and reality in students' daily lives

Francileide Martins P. de Sá Leitão

Statistics and reality in students' daily lives

ScienciaScripts

Imprint

Any brand names and product names mentioned in this book are subject to trademark, brand or patent protection and are trademarks or registered trademarks of their respective holders. The use of brand names, product names, common names, trade names, product descriptions etc. even without a particular marking in this work is in no way to be construed to mean that such names may be regarded as unrestricted in respect of trademark and brand protection legislation and could thus be used by anyone.

Cover image: www.ingimage.com

This book is a translation from the original published under ISBN 978-613-9-65218-1.

Publisher:
Sciencia Scripts
is a trademark of
Dodo Books Indian Ocean Ltd. and OmniScriptum S.R.L publishing group

120 High Road, East Finchley, London, N2 9ED, United Kingdom
Str. Armeneasca 28/1, office 1, Chisinau MD-2012, Republic of Moldova, Europe
Printed at: see last page
ISBN: 978-620-7-86155-2

DEDICATORY

I dedicate it to Flaviano, my beloved husband, to my sons Pedro Arthur and Flaviano Filho and to my parents Francisco Pinheiro and Maria de Lourdes, for their love, encouragement and understanding during this stage in my life, in which I sought to acquire knowledge for my profession.

ACKNOWLEDGEMENT

To God, in the first place, who has always been by my side.

To my husband, for the companionship and generosity with which he took on some of my undemanding family duties throughout my master's degree.

To my sons Pedro Arthur and Flaviano Filho for the sincere and pure love that makes life more rewarding.

To my parents and my sister Fabiola, for their dedication and affection during this period of professional growth.

To my advisor, Prof Dr Antônio Ronaldo Garcia Gomes, for his support in this endeavour, for his excellent guidance, competence, availability and friendship.

To the headmasters and coordinators of the schools where I teach, and especially to the 9° grade students of the Santo André Educational Complex who collaborated as research subjects for this dissertation.

The coordination and all the teachers at the Federal Rural University of the Semi-Arid Region (UFERSA), for sharing their knowledge.

"Men make their history, but they don't make it the way they want, they don't make it under circumstances of their own choosing, but under those with which they are directly confronted, bequeathed and transmitted in the past." (The Eighteenth Brumaire of Louis Bonaparte).

Karl Marx, 1996

SUMMARY

Statistics is a part of applied mathematics that studies methods for collecting, organising, analysing and interpreting data, since its study is aimed, among other things, at decision-making. So we have to experience it so that we can understand it better without skipping any steps. In this way we will better interpret and consequently improve skills and competences which, although important, should not just be measured by a written test, but by an application of a concept that has been constructed and experienced by the student in everyday life. Statistics therefore seeks situations in which problem-solving is meaningful for the student and mobilises their cognitive resources.

According to the PCN's (2002), the teaching of mathematics can help students develop skills related to representation, comprehension and investigation, as well as socio-cultural contextualisation. This means placing students in a teaching-learning process that values mathematical (statistical) reasoning and also enables the teacher to work on content in an interdisciplinary way and broaden the understanding of its meanings, thus contributing to the formation and construction of knowledge.

KEYWORDS: Statistics, everyday life, maths, collection and interpretation.

SUMMARY

CHAPTER 1

INTRODUCTION

Statistics is now a part of school curricula and has been gaining momentum as the ENEM approach has grown, given the need for statistical knowledge in our daily lives. A large proportion of the media (magazines, newspapers, radio, television and the internet) use statistical models such as graphs, tables and pictograms to enrich the information they disseminate to the public.

A large part of this audience ends up not deciphering this complex quantitative and qualitative language.

This change in perception, which takes place in the problematisation of a conflicting reality, implies a new confrontation between individuals and their reality. It implies an "appropriation" of the context, an insertion into it, no longer being "adhered" to it; no longer being almost under time, but in it. FREIRE (1981).

Statistics is a science that deals with the collection and processing of information. Its aim is to analyse the data collected, describing and organising it for later interpretation and possible use in predicting future events. As soon as the rulers of the first civilisations had knowledge of the state's assets and how they were distributed among the inhabitants, the first statistics were made, particularly to determine tax laws. Hence the name statistics comes from the term "state".

The statistical sampling process generates information that helps to make projections, answering important questions in business, organisation and government activities.

Statistics as a science only came about in the 18th century, in the records of the German Godofredo Achenwall, still as a non-regular cataloguing of data (CRESPO 2009).

The National Curriculum Parameters (PCN) emphasise the curricular aspect of statistics:

In a world where social, cultural and professional needs are taking on new contours, all areas require some competence in maths and the ability to understand mathematical concepts and procedures is necessary both for drawing conclusions and making arguments, and for citizens to act as prudent consumers or make decisions in their personal and professional lives. Maths in secondary school has a formative value, helping to structure thinking and deductive reasoning, but it also plays an instrumental role, as it is a tool used in everyday life and for many specific tasks in almost all human activities (BRASIL, 1999).

In this way, the PCN emphasises a comprehensive education with regard to the quantitative education of students in primary school.

It is of fundamental importance that the practices and content taught in the classroom are in tune with the new demands of the world we live in, so that education is not something distant from the daily lives of our students, but is an integral part of their experiences for a better existence. Stimulating the ability to read and interpret graphs and facts is the role of school work in the search for information for a fully-fledged citizen.

CHAPTER 2

HISTORY OF STATISTICS

2.1 The emergence of statistics

Statistics is one of the sciences whose first steps date back to the dawn of human history and whose formal development tends to be in tune with the evolution of human knowledge. It is also a science that leaves no doubt as to its predisposition to incorporate techniques, discoveries and new theories, whether its own or from other specialities, as well as underpinning new discoveries in multiple areas of knowledge.

Since our ancestors, we have lived in a world where most people consider maths to be a boring and very difficult subject, so these qualities are also attributed to statistics, as it is part of mathematics.

Its origins can be traced back to the need to organise information and registers for the state, according to its forerunners the German Conring (1660) and the English John Graunt (1962), Willian Petty (1982) and Halley (1694). Graunt published a study of the records of baptisms, marriages and burials that had been kept in parishes for a century. Petty, creator of the term "Political Arithmetic", was the first to make conjectures based on statistical information, using tables and relative numbers. Halley noticed that death, which was very irregular and unpredictable for individual cases, followed a reasonably fixed law if a large number of people were counted. In 1708, the University of Lena inaugurated a course in Statistics and later, in 1749, Godofred Achenwal, at the University of Gottingen in Germany, generalised the name "Statistics", which is still accepted today, defining its object and its relationship with the sciences.

The so-called "technical and scientific improvement" phase began in 1853 with the first Statistical Congress. From then until the present day, it has been used more and more intelligently in various fields, and the process of statistical elaboration has been increasingly perfected.

Statistics is as old as the first man, because there has always been a need to enumerate things. From its beginnings as a simple compilation of numbers, statistics has evolved to the present day as a powerful tool for researching phenomena and coincidences. Educators, politicians, economists, doctors, farmers, industrialists and others all use statistics in their respective fields of work.

2.2 Applications of Statistics

We know that since its inception, statistics has had countless applications and that it is becoming more and more indispensable today. In Meteorology, it is important for the maritime and aerial areas, as it gives the coordinates of pressures, temperatures, humidity, rainfall, winds and others that are found through statistical data. For farmers, it provides guidance on harvests, financial yields, etc.

The geographer arrives at his conclusions about migratory currents, climate and density through statistical data. Merchants study trends in order to buy and sell their goods. The number of inhabitants, deaths and voters is found through statistical research. In elections, sample surveys are widely used to get an idea of the whole and from there to get an idea of who will win, as well as to encourage voters to vote for a candidate or political party.

This statistical data influences a result depending on the data collected, because as the complexity of a certain fact grows, it becomes more difficult and time-consuming to reveal its fundamental law.

Statistics can be seen in its applications as a method of discovering relationships of dependence between causes and effects of any phenomenon, a logical instrument, based on an inductive method, aimed at discovery for the betterment of humanity.

Statistical information is effective, concise and specific, thus providing extremely important input for rational decision-making. It provides tools for institutions and companies to define their goals, identify their problems and perform better in today's competitive market.

Statistics is a science that serves others and is an interdisciplinary subject. It is present in various professions and situations such as: administration, market research, the humanities, biometric sciences (agronomy, psychology, medicine, etc.), the financial market, planning, budgeting, sales forecasts, etc. This shows how interdisciplinary it is, as it offers great hope for renewal and change. It provides the necessary conditions for a dialogue between disciplines to coexist. Its purpose is to establish a relationship that leads students to understand, process, think about, criticise and incorporate the different contents and the links between the disciplines, allowing them to build a coherent and logical construction of the knowledge acquired in the different areas.

> We believe that the implementation of an interdisciplinary practice by teachers involves, among other things, knowledge of the concept of interdisciplinarity, mastery of the content of the disciplines and the relationship between the actors involved in school practice (CRUSOÉ, 2009, p.22).

Interdisciplinarity and statistics are therefore very important in pedagogical practice, as they go beyond knowledge, as they guarantee a high level of maturity for those who practise them. In this process, we must make students aware that they are active, committed and responsible agents for interacting in the environment in which they live, thus contributing to improving the teaching-learning process.

At the Santo André Educational Complex school in the city of Assú, where I teach, I have an interdisciplinary project with the art and writing teachers. We make a magazine about geometric solids, so I explain the content, the writing teacher corrects the text and the art teacher builds the solids together with them, then the students take it to the printer who reproduces it and we present it at school. In this way, students revise content in different classes, which significantly improves their learning.

CHAPTER 3

BASIC CONCEPTS OF STATISTICS

3.1 The Importance of Statistics as a Science

Since ancient times, various peoples have recorded the number of inhabitants, births and deaths, made estimates of individual social wealth, equitably distributed land to the people, levied taxes and carried out quantitative surveys using processes that today we would call "statistics".

Statistics is a very important subject, as it involves communication. From the 6th grade onwards° , teachers can already work with these concepts, since they are present in people's daily lives.

Most news stories in newspapers and magazines use graphs and tables to express information of various kinds, because they are easy to understand. The responsible use of research results is important for governments, companies and even individuals to direct their actions towards the interests of the community.

Most of the time, a statistical survey is carried out using a sample, where we consult a representative group of people; another way is to use the population or research universe, when we consult the whole.

According to the PCN's (2001, p. 29), it is important that mathematics plays a balanced and inseparable role in the formation of intellectual capacities, in the structure of thought, in speeding up the student's deductive reasoning, in its application to problems, everyday life situations and activities in the world of work, and in supporting the construction of knowledge in other curricular areas.

Statistical studies are currently advancing rapidly and, with their processes and techniques, have contributed to the organisation of business and resources in the modern world. It is therefore important to be able to read a table, interpret a graph and make estimates.

It is therefore necessary that, from primary school onwards, students have contact with this branch of mathematics that is so important and used in their daily lives, learning how to collect data, organise it in tables, construct graphs and calculate averages. These skills will help them in various mathematical calculations, as estimating a result can prevent many errors, especially those of data interpretation.

We don't think it's easy to develop this content, but we believe it's no longer possible for students to reach secondary school without having this knowledge, which is so common in the media. Information processing is a current language that needs to be known and discussed. It is also very important for students to realise that research does not represent absolute truths, but rather trends, when properly carried out. However, starting with data collection, making tables, constructing graphs and interpreting these results, a lot of care needs to be taken, as trends may or may not be confirmed in reality.

In this way, everyone should be able to understand this information, as Imenes and Lellis (1994, p.11) state when dealing with this relationship:

> In modern societies, much of the information is conveyed in mathematical language. We live in a world of percentage rates, multiplicative coefficients, diagrams, graphs and statistical truths. To decode this kind of information, you need information, you need maths instruction. Here we see a first relationship between maths teaching and the conditions necessary for exercising citizenship.

In this way, we will provide students with good challenges to encourage them to develop strategies for solving problems. In this way, we aim to show students that statistics is present in their everyday lives.

GENERAL OBJECTIVE

To help integrate students into the society in which they live by providing them with basic knowledge of the theory and practice of maths, stimulating their curiosity, interest and creativity so that they can explore new ideas and discover new ways of applying the concepts they have learnt and interpreting and solving problems.

SPECIFIC OBJECTIVES

- Recognise quantitative and qualitative variables;
- Draw up and interpret absolute and relative frequency tables;
- Draw up and interpret bar, column, segment and sector graphs;
- Calculate and interpret measures of central tendency: mean, median and mode.

3.2   Statistics taught in primary school

Statistics is a part of applied mathematics that provides methods for collecting, organising, describing, analysing and interpreting data and using it to make decisions.

The collection, organisation and description of data is the responsibility of Descriptive Statistics, while the analysis and interpretation of this data is the responsibility of Inductive Statistics.

Collection can be direct or indirect.

Direct collection is made of compulsory records such as births, deaths, marriages and others. It can be Continuous when it is done continuously such as student attendance in class, Periodic when it is done at constant intervals such as monthly student evaluations and Occasional when it is done to meet an emergency such as an epidemic.

Indirect data collection uses data from direct data collection or phenomena that have already been studied. For example, infant mortality.

By collecting data we can get to know the geographical and social reality, the natural, human and financial resources, the expectations of a community or a company, and set goals so that its objectives are achieved in the short, medium or long term.

Statistics creates the possibility of strategies to be adopted in an enterprise, in the choice of techniques for verifying and evaluating quantity and quality, as well as profits and losses.

A population is a group of people, objects or events about which we want to obtain information. For example: In an elementary school, a survey was carried out with all the students.

A sample is a representation of the population as a whole. Taking the previous example, we have: The school has 1500 students from which we randomly

180 were chosen to take part in the survey. So 1500 is the population and 180 is the sample.

So the aim of statistics is to draw conclusions about the whole (population) from the information found in the representative part (sample).

Statistical Variable

Qualitative variables are those that classify elements in relation to attributes or qualities. Example: A survey of 6th graders about their favourite break time activity gave the following results: talking, snacking, playing and reading.

Quantitative variables are those that indicate a number as a result of counting or measuring. Example: A survey was carried out in the 7th grade to find out who was in the grade and the results were: 11 years old, 12 years old, 13 years old, 14 years old and 15 years old.

Frequencies

Absolute Frequency is the number of times a variable value is mentioned in a survey. For example: Let's toss a coin 8 times and look at the results.

Table 1: Absolute Frequency

CURRENCY	ABSOLUTE FREQUENCY
Cara	3
Crown	5

The frequency with which the crown appeared was greater than that of the face.

Relative Frequency is the ratio between the absolute frequency of a variable and the total number of mentions of all the variables in the survey.

Using the example above, the relative frequency can be represented as a fraction or as a percentage.

Table 2: Absolute and Relative Frequencies

CURRENCY	F. ABSOLUTE	F. RELATIVE	F. RELATIVE
Cara	3	3/8	0,375=37,5%
Crown	5	5/8	0,625=62,5%
Total	8	1	100%

CHART TYPES

Sector or Pie charts - used to emphasise the participation of the data in the survey in relation to the whole surveyed and generally used in percentage terms.

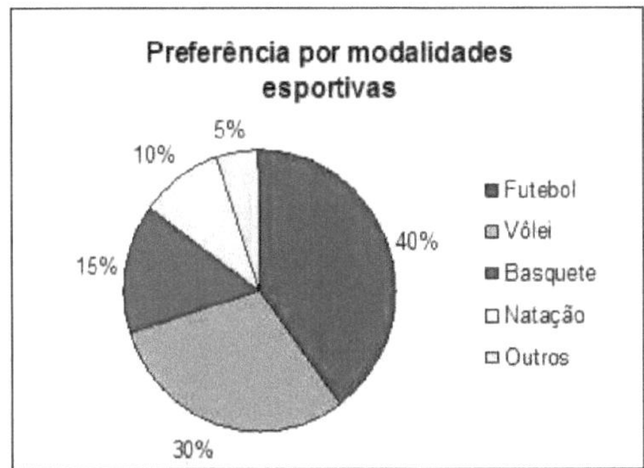

Graph 1: Preference for sports between the ages of 9 °

Segment or Line Graphs - are generally used to show the evolution of the frequencies of the values of a variable over a certain period of time.

Graph 2: Job vacancies for managers and direct employees in São Paulo

Source: São Paulo Newspapers - 30/09/2005

Bar Graphs - are used to compare the frequency of values of the same variable at a given time.

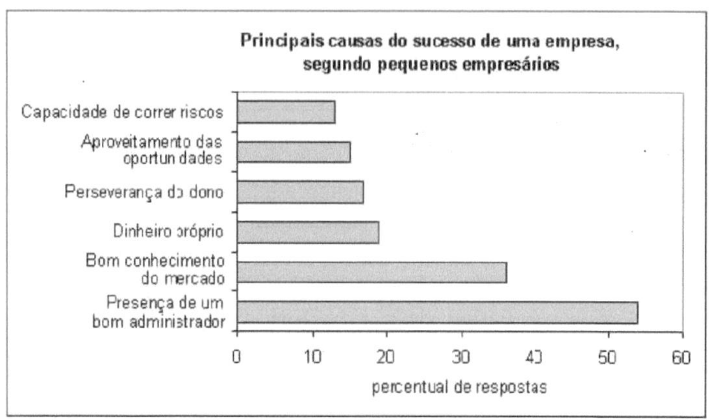

Graph 3: The main reasons for a company's success, according to small business owners.

Source: http://www.luqli.com.br/2008/02/grafico-de-barras

Column chart - this is similar to the bar chart in terms of employment.

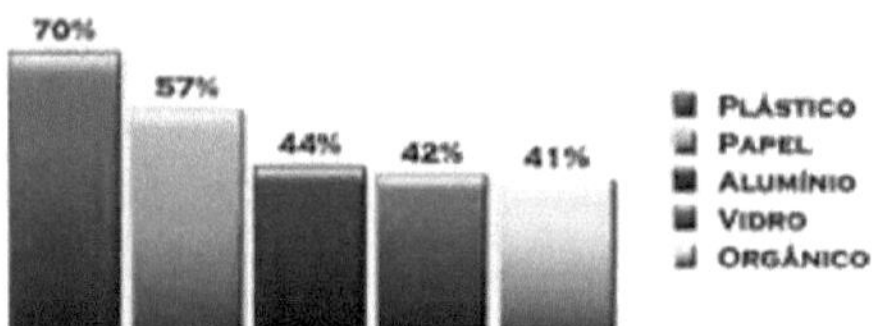

Graph 4: Materials sorted at home

Source: http://reciclagemcti.blogspot.com.br/2009_11_01_archive.html

Histogram - is used in surveys where the variable is continuous, whose values are indicated by classes (intervals). The histogram is made up of juxtaposed rectangles, the bases of which are located on the horizontal axis.

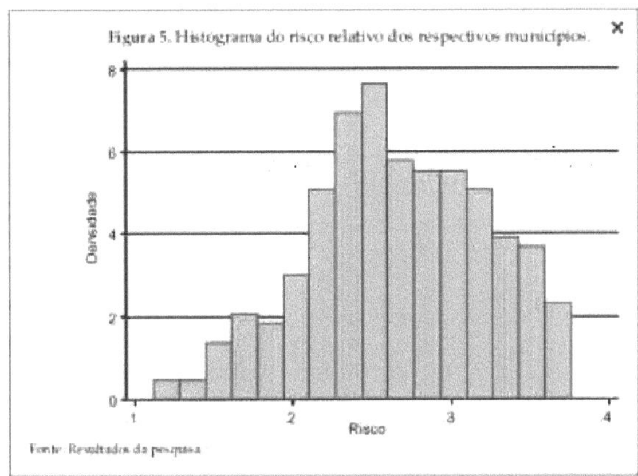

Graph 5: Example histogram

Source: www.scielo. br/scielo. ph p?pid=S0103- 20032009000300001 &script=sci_arttext

Pictogram - is represented by figures that translate information into drawing form.

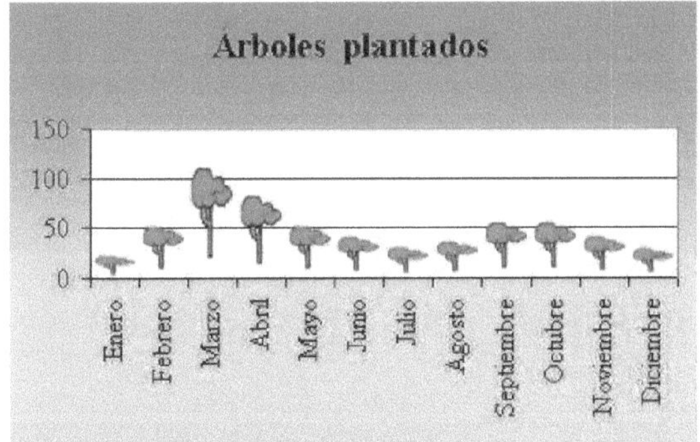

Graph 6: Trees planted during 1 year
Source:

http://www.ceibal.edu.uy/UserFiles/P0001/ODEA/ORIGINAL/090316_estadistica.elp/pi

ctograma.html

MEASURES OF CENTRAL TENDENCY

Several grades taken by a student over the course of a year, for example, can be represented by just one grade.

This is a situation involving trend measures. The measures of tendency are: Mean, Mode and Median.

1-   Averages

a)   Arithmetic - the ratio of the sum of the variable's values to their number.

$$M_a = \frac{x_1 + x_2 + x_3 + \cdots + x_n}{N}$$

Example: Consider 5 classmates who study at the same school. Their heights are as follows: 1.75m; 1.60m; 1.35m; 1.20m and 1.18m. What is the average height of these classmates?

$$M_a = \frac{1,75 + 1,60 + 1,35 + 1,20 + 1,18}{5} = \frac{7,08}{5} = 1,42$$

b) Weighted - the ratio between the sum of the products (frequency x value of the variable) and the total of the frequencies.

$$M_p = \frac{p_1 \cdot x_1 + p_2 \cdot x_2 + p_3 \cdot x_3 + \cdots + p_n \cdot x_n}{p_1 + p_2 + p_3 + \cdots + p_n}$$

Example: the survey of maths grades in a class of 9° has been summarised in the table.

Table 3: Frequency of grades in Year 9 °

No. of students	Note
2	3,0
3	4,0
10	5,0
15	6,0
10	7,0
6	8,0
4	9,0
3	10,0

Source: Author using Word-2007

What is the average grade in this class?

$$M_p = \frac{2 \cdot 3{,}0 + 3 \cdot 4{,}0 + 10 \cdot 5{,}0 + 15 \cdot 6{,}0 + 10 \cdot 7{,}0 + 6 \cdot 8{,}0 + 4 \cdot 9{,}0 + 3 \cdot 10{,}0}{2 + 3 + 10 + 15 + 10 + 6 + 4 + 3}$$

$$M_p = \frac{6 + 12 + 60 + 90 + 70 + 48 + 36 + 30}{53} = \frac{342}{53} = 6{,}45$$

2-   Mode - is an expression used to designate the highest frequency within a statistical survey.

Example: Consider the group of values and identify the mode.

a)   2 reals, 1 real, 2 reals, 1 real, 2 reals and 2 reals.

The fashion is 2 reais

b)   4 reais, 6 reais, 8 reais, 8 reais, 10 reais, 12 reais, 8 reais, 14 reais, 18 reais, 14 reais and 14 reais.

The mode is 8 and 14, which is called bimodal.

3- Median - is a number or an arithmetic mean that is arranged in a certain order.

Example: The ages of the six people in Luana's family in years are: 6, 41, 17, 14, 12 and 42, what is the median?

6-12-14-17-41-42

$$M = \frac{14 + 17}{2} = \frac{31}{2} = 15{,}5$$

Example: The number of goals scored by Brazil in the 2006 World Cup were: 0,1, 0, 0, 0, 1,2, 1, 2, 1, 1, 3, 1, 1, 3, 3, 4, 4, 5, 5, 3, 3.

0-0-0-0-1-1-1-1-1-1-1-1-2-2-3-3-3-3-3-4-4-Õ-5

The median is 1

3.3 Applied research methodologies

The main focus of this research is to show how statistics is present in everyday life. The methodology used to investigate these issues is a qualitative and quantitative approach of a descriptive and explanatory nature, supported by field research. The qualitative approach is being widely used as a research methodology in education and expresses the best complexity for human and social phenomena.

The descriptive characteristics reveal the type of research methodology developed here, showing how statistics is growing every day, especially in the education of citizens, as it provides tools that allow human beings to develop strategies, face challenges, prove and justify results, among other activities. In addition, it stimulates creativity, the development of reasoning and collective work, which is of the utmost importance for the formation of the student as an individual.Collective work favours interaction between all the students belonging to the same group. This gives them the chance to put forward ideas, argue about their points of view and discuss different strategies and solutions. It is important for the teacher to pay attention to the way the students are organised in a given activity, which should allow them to satisfactorily achieve the objectives set for the activity.

Figure 1: Students studying the data collected
Source: Author using digital camera

If statistical work is to be carried out properly, it must go through the following stages:

a) Data collection

b) Data collation

c) Display of results

d) Interpreting the facts

A statistical survey is one of the essential preliminary operations. In addition to the usual means, such as publications in printed newspapers, posters, projections and illuminated adverts, postcards and stamps, the use of radio and aeroplanes, advertising should be carried out in schools, so that students induce their parents to answer the questions well.

A questionnaire survey consists of a series of questions and blanks for answers. The questionnaire can be filled in by the researcher or the interviewee, but it has to be very clear in order to avoid ambiguous answers.Some people refuse to answer or answer poorly to a questionnaire for the following reasons:

a) Lack of clarity in the questionnaire

b) Misinterpreted questions

c) Ignorance

d) Doubt about the release of data

e) Fear of tax increases

f) Represents

These questionnaires are then used to collate the data, where they are usually transformed into tables so that the results can be better understood. When the amount of information is large, we realise that manual counting is difficult, forcing us to resort to computerised counting. An example is the counting of elections, because in the past it took a long time and today with computerisation we have the result on the same day.The activity in Annex I was carried out in the classroom and it can be seen that the students did it calmly, because it was an engaging activity where in the fourth question they had to interact.

Figure 2: Students constructing a frequency table.
Source: Author using digital camera

Annex II begins the activity applied to students in the 9th° year of primary school. The questionnaire, made up of questions and applied in the noble neighbourhoods of Janduís and Novo Horizonte, provides a sample of the level of education of the population of Assuense.

According to Gil (2002, p.115), an interview is a technique "involving two people in a face-to-face situation" in which one of them asks questions and the other responds.

Annex III With the information in hand, the next stage was to organise the data collected using the questionnaire in Annex II. Then organise it in the form of an absolute and relative frequency table.

Figure 3: Student constructing a pie chart
Source: Author using digital camera

By analysing and interpreting this data, the students calculated percentages and angles (pie charts) and demonstrated the results of this research.

We can highlight some important aspects in relation to the characterisation of those involved in this research as follows: the vast majority of those interviewed did not complete primary school and this is worrying, because we live in a globalised world today that has various ways of completing our studies. There are supplementary schools for primary and secondary education, not to mention distance learning, and what is lacking for these people to finish? Lack of incentive, especially from family members, having to start work at a young age, or perhaps often a lack of interest.

3.4 Evaluation aspects

Through problem situations in which the real context of the student's everyday life is taken into account. Where the organisation of tables and the information on the subject encourages students to observe and make comparisons about the situation or phenomenon in question. Consequently, it favours the development of their ability to estimate, express opinions and make decisions.

In this way, assessment is becoming a fundamental tool for providing information on how the teaching-learning process as a whole is going. It provides a continuous and dynamic diagnosis of the process, allowing us to rethink and reformulate teaching methods, procedures and strategies so that students really learn.

"Evaluating learning, therefore, implies evaluating the teaching offered, for example, if there is no expected learning, it means that the teaching did not fulfil its purpose: to make people learn." (PCN's vol. 1- Brasília, 1997)

In order to continuously assess students, teachers can use a variety of resources, such as lists of activities, assignments with seminar presentations, reports and written tests, among others. They need to be given the opportunity to check their difficulties and needs in constructing knowledge. And through assessment, they can become aware of the content they have already learnt and also identify whether they need to dedicate more time to certain subjects.

CHAPTER 4

FINAL CONSIDERATIONS

We saw that there is a diversity of learning, as some of the students are very participative and others are very introspective. I found it difficult to get them into the field, in other words, to collect data, because they were students in the 9th year of primary school, underage, so if anything happened to them it would be entirely my responsibility. According to them, some people refused to answer and others treated them badly.

Although the school was private, it didn't have a computer lab, so they brought their laptops into the classroom so that we could build the tables and graphs. They were excellent at understanding the content, especially when it came to drawing graphs, identifying the mode, median and calculating the mean.

Initially I thought I'd do this research at the JK school where I teach, but I work at night and my students work during the day, which made it impossible, as they had no way of taking time off work. In most of the classrooms, the students are out of age, for example, I have students who are old enough to be my grandparents who are already retired and are just about to get their secondary school leaving certificate.

The educator, in turn, must articulate a teaching dynamic geared towards motivation, creating activities that include content that is appropriate to the students' experiences. These activities help to develop not only mathematical reasoning, but also students' social and emotional skills. So I realised that the teaching staff were more involved than last year, because the activities they experience attract more attention, as well as learning what is in the curriculum, the students review the content and renew teaching methods.

Teachers should seek to bring students into situations where they experience and better understand the content being covered, as this promotes student growth and maturity. The vision of the applicability of mathematical concepts from a broader perspective, inserting the student's knowledge into real, everyday life.

Students need to express their ideas through writing or through dialogue with the teacher and classmates. By interacting with their peers during certain activities, they have the opportunity to develop their ability to organise their reasoning, to argue for it and to listen to their peers. This contributes to the development of mutual respect, co-operation, critical thinking and more.

We live in a world that is bombarded with information. We therefore need to know how to select, understand, interpret and form our own opinions in order to be participative and active citizens in the society in which we live. The media appropriates information of all kinds, often conveyed through graphs, spreadsheets and tables. These tools have become commonplace due to the easy visualisation and understanding of statistical data.

In this context, the educator's role is to select and discuss relevant subjects in the classroom, using information that is strictly linked to reality as an ally in teaching. In this way, it's possible to link curriculum content with young people's everyday lives.

We must encourage and motivate our students to learn and realise the importance of statistics in primary school, secondary school, the Enem itself and, above all, in their daily lives.

We have to try to stimulate curiosity so that they socialise and contextualise arithmetic, algebra, geometry, financial mathematics and statistics with reality. The teaching and use of statistical models in the classroom must be in line with the needs, interests and life experiences of our students. According to LOPES (2003):

> Statistics is a science that is not restricted to a set of techniques. It
> provides knowledge that allows us to deal with uncertainties and the

variability of data, even during data collection, making it possible to take decisions with better arguments.

It can be concluded from this that the teaching of statistics is one of the fundamental elements for the social and intellectual formation of students, making them knowledgeable individuals and citizens prepared to be part of a society full of changes. It should be emphasised that it is essential to develop autonomy, criticality, the ability to argue and creativity, thus proving the importance of statistics in the life of the citizen.

CHAPTER 5

REFERENCES

BOLFARINE, Heleno; SANDOVAL, Mônica Carneiro. Introduction to Statistical Inference. Applied Maths Collection. Editora SBM.

BRAZIL, Ministry of Education. National Curriculum Parameters: Secondary Education. Brasília: Ministry of Education, 1999.

Ministry of Education. Secretary of Secondary and Technological Education. National Curriculum Parameters. Secondary Education. Brasilia-2002.

National Curriculum Parameters, volume 1- Brasília: SEF/ MEC, 1997. Introduction.

National Curriculum Parameters: Ministry of Education. Department of Basic Education. 3rd edition - Brasília. The Secretary, 2001.

CRESPO, Antônio Arnot. Statistics Made Easy. Editora Saraiva 19ª updated edition, 2009.

CRUSOÉ, Nilma Margarida de Castro. Interdisciplinarity: social representations of maths teachers. Natal, RN: EDUFRN- Editora de UFRN, 2009.

FREIRE, Paulo. Cultural action for freedom. Rio de Janeiro: Paz e Terra, 1981.

GIL, A. C. Como Elaborar Projetos de Pesquisa. São Paulo: Atlas, 2002 p.42 and 120.

IEZZI, Gelson; DOLCE, Osvaldo; MACHADO, Antônio. Maths and Reality. Volume: 4. 9° year Editora Atual- 2009.

IMENES, Luiz Márcio; LELLIS, Marcelo. Maths teaching and the formation of citizens. Maths Education: Philosophical Foundations and Social Challenges.

Temas e Debates, Rio de Janeiro, 1994.

LEVIN, J. Statistics Applied to the Human Sciences. São Paulo, Editora Harbra, 1987.

LOPES, C. A. E. Teachers' professional knowledge and its relationship with statistics and probability in early childhood education. 2003. 281 pages PhD Thesis in Maths Education - State University of Campinas, Campinas-SP.

MARX K. Practice, History and the Construction of Knowledge. In: Andery, M. A. p. A. et. al. Para Compreender a Ciência. São Paulo/Rio de Janeiro, 1996.

MILONE, Giuseppe. General and Applied Statistics. Editora Cengage Learning, 2009. 2ª edition.

MORI, Iracema; ONAGA, Dulce Sativo. Maths Ideas and Challenges. 9th year Editora Saraiva-2009-SP. Reformulated 15th edition.

SEVERINO, Antônio Joaquim. Metodologia do Trabalho Cientifica. 23rd edition revised and updated - São Paulo: Cortez, 2007.

Access to the Digital Library of Theses and Dissertations at 31

Available at www.mec.gov.br (accessed on 05/01/13)

Available at www.portalpositivo.com.br (accessed on 19/12/12)

Available at www.ser.com.br (accessed on 27/12/12)

CHAPTER 6

ANNEX I

Proposed exercise

1- The table below is the result of a survey of the best-selling "music genres" in a CD shop over the course of a month. Note that the table is incomplete and try to complete it.

MUSICAL GENRE	F. ABSOLUTE	F. RELATIVE
Country		35%
MPB	48	
Rock		
Classic		10%
Total	200	

2- What type of variable does the above question refer to?

3- A group of students were asked which football team they preferred. The votes were recorded as follows (Brazil was not included).

Germany llllll   Holland lllllllll   Spain 1
Argentina lll   Italy lllllll
England ll   France llll

Construct a frequency table corresponding to this survey.

4- Make a "replica" of this research in your classroom. Then construct the frequency table.

5- Consider a group made up of five friends aged 13, 13, 14, 14 and 15. What happens to the average age of this group if a sixth friend aged 16 joins the group?

a) It remains the same

b) Decrease by one year

c) Increase by 12 years

d) Increases over a year

e) Increases in less than a year

6- You did two assignments in one term and got marks of 8.5 and 5.5. What grade do you have to get on the 3rd° assignment for the average of the three assignments to be 7?

7- The cost of feeding a family for a week during the July holidays was:

DAYS	VALUES
Sunday	28,00
Monday	32,00
Tuesday	32,00
Wednesday	40,00
Thursday	26,00
Friday	24,00
Saturday	50,00

Calculate:

a) The daily average

b) The trend of spending

c) Median spending

ANNEX II

PROPOSED ACTIVITY

Objectives:

- Collect data for statistical study;

- Analysing the data collected.

Material:

- Interview sheet prepared by the teacher

- Ruler

- Compass

- Calculator

Model sheet:

A sample of the educational level of three streets in a neighbourhood.

1-What is your name?_______________________________

2- How old are you? _______________________________

3-Street: _______________________________________

4-Bai rro: ______________________________________

5-What is your level of education?

   a) Illiterate
   b) Incomplete primary education
   c) Complete primary education
   d) Secondary school incomplete

   Completed high school

   e) Incomplete higher education
   f) Higher education completed

FREQUENCY TABLE

GRADE	ABSOLUTE FREQUENCY	RELATIVE FREQUENCY
Illiterate		
Incomplete Primary Education		
Complete Primary Education		
Incomplete Medical Education		
Complete medical education		
Complete secondary		

education		
Complete Higher Education		
Total		

Calculations:

Sector Charts:

Annex III

The student body is working in class with the data collected in the Novo Horizonte and Janduís neighbourhoods. With the teacher's help, they are constructing absolute and relative frequency tables and then calculating the percentage data to be presented in a pie chart.

Figure 4: Students constructing a pie chart of the field research Source: Author using digital camera

The 9th grade class° making absolute and relative frequency tables for the Novo Horizonte and Janduís neighbourhoods in the city of Assú with the teacher's help.

Figure 5: Teacher helping studentsSource: Author using digital camera

The students in Year 9° constructed sector graphs with a protractor and then checked the results.

Figure 6: Students checking the pie chart

Source: Author using digital camera
The teacher stops by the groups to check on their progress and help them if necessary.

Figure 7: Teacher observing students at work

Source: Author using digital camera
The student body is calculating the arithmetic mean, mode and median of the survey results.

Figure 8: Students calculating mean, median and mode

Source: Author using digital camera
The student body organising their work and then uploading it to the computer.

Figure 9: Students completing the activity

Source: Author using digital camera

Annex IV

Table 4: Frequency of the Novo Horizonte neighbourhood

Level of Education	F.A	F.R
Illiterate	10	$\frac{10}{192} = 5\%$
Incomplete primary education	58	$\frac{58}{192} = 30\%$
Complete primary education	9	$\frac{9}{192} = 5\%$
Secondary school incomplete	26	$\frac{26}{192} = 14\%$
Completed high school	43	$\frac{43}{192} = 22\%$
Incomplete university degree	13	$\frac{13}{192} = 7\%$
University degree completed	33	$\frac{33}{192} = 17\%$
Total	192	100%

Source: Year 9 students .°

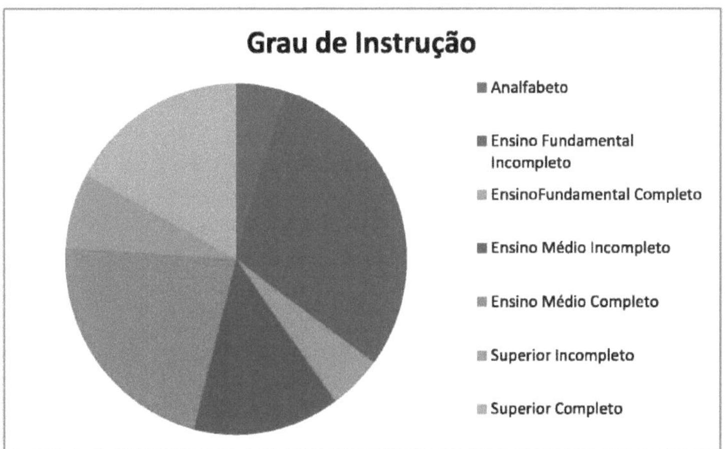

Graph 7: Graph of Sectors in the Novo Horizonte neighbourhood
Source: Year 9 students °

Table 5: Frequency of the Janduís neighbourhood

Level of Education	F.A	F.R
Illiterate	6	$\dfrac{6}{181} = 3\ \%$
Incomplete primary education	49	$\dfrac{49}{181} \cong 28\%$
Complete primary education	9	$\dfrac{9}{181} \cong 5\%$
Secondary school incomplete	19	$\dfrac{19}{181} = 10\%$
Completed high school	55	$\dfrac{55}{181} = 30\%$
Incomplete university degree	12	$\dfrac{13}{181} \cong 7\%$
University degree completed	31	$\dfrac{31}{181} = 17\%$
Total	181	100%

Source: Year 9 students .°

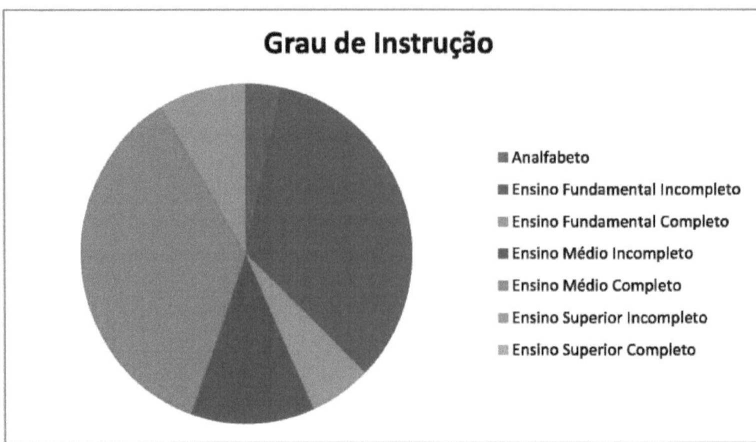

Graph 8: Sector graph for the Janduís neighbourhood.

Source: Year 9 students °

Printed by Books on Demand GmbH, Norderstedt / Germany